JOURNEY THROUGH OUR SOLAR SYSTEM

EXPLORING COMETS AND ASTEROIDS

By David Anthony

KidHaven PUBLISHING

Published in 2018 by
KidHaven Publishing, an Imprint of Greenhaven Publishing, LLC
353 3rd Avenue
Suite 255
New York, NY 10010

Designer: Deanna Paternostro
Editor: Vanessa Oswald

Photo credits: Cover solarseven/Shutterstock.com; back cover, p. 15 Vadim Sadovski/Shutterstock.com; p. 5 Christos Georghiou/Shutterstock.com; p. 7 Triff/Shutterstock.com; pp. 8–9 paulista/Shutterstock.com; p. 11 muratart/Shutterstock.com; p. 13 Digital Vision./Getty Images; p. 17 Andrea Danti/Shutterstock.com; p. 19 Mopic/Shutterstock.com; p. 21 Getty Images/Stringer/Hulton Archive/Getty Images.

Cataloging-in-Publication Data

Names: Anthony, David.
Title: Exploring comets and asteroids / David Anthony.
Description: New York : KidHaven Publishing, 2018. | Series: Journey through our solar system | Includes index.
Identifiers: ISBN 9781534522947 (pbk.) | 9781534522756 (library bound) | ISBN 9781534522480 (6 pack) | ISBN 9781534522688 (ebook)
Subjects: LCSH: Comets–Juvenile literature. | Asteroids–Juvenile literature.
Classification: LCC QB721.5 A58 2018 | DDC 523.5'1–dc23

Printed in the United States of America

CPSIA compliance information: Batch #BS17KL: For further information contact Greenhaven Publishing LLC, New York, New York at 1-844-317-7404.

Please visit our website, www.greenhavenpublishing.com. For a free color catalog of all our high-quality books, call toll free 1-844-317-7404 or fax 1-844-317-7405.

CONTENTS

ROCKS IN OUTER SPACE

Many planets and moons make up parts of the **solar system.** Other objects that travel through the solar system include comets and asteroids.

Craters on planets and moons in the solar system are formed by comets and asteroids crashing into them.

FLYING COMETS

Comets are flying chunks of icy rock in space. They are made in a very cold area far from Earth. Comets are sometimes called "dirty snowballs."

Scientists believe comets are left over from what formed our solar system.

The **energy** from the sun melts some of the ice from the comets when they pass by it.

Gottfried Kirch discovered the first comet by **telescope** in 1680. It was named the Great Comet of 1680 and was known for its very long tail.

Gas and dust are released from comets when they get near the sun. This makes the comets look like they have tails.

It takes many years
for comets to travel
around the solar system.
Comets **orbit** the sun in
the same way planets do.

As of 2010, about 4,000 comets have been discovered.

HALLEY'S COMET

Some comets pass through the solar system and are never seen again. Some comets, such as Halley's comet, have been seen from Earth more than once. Halley's comet can be seen in the night sky every 75 to 76 years.

Shown here is Halley's comet.

AMAZING ASTEROIDS

Asteroids are rocks in space left over from the birth of the solar system. They can be miles long or as small as grains of sand.

Some scientists believe an asteroid may have wiped out the dinosaurs 65 million years ago!

The asteroid belt is where most asteroids orbit the sun. This area is between Mars and Jupiter. There are millions of asteroids in the asteroid belt.

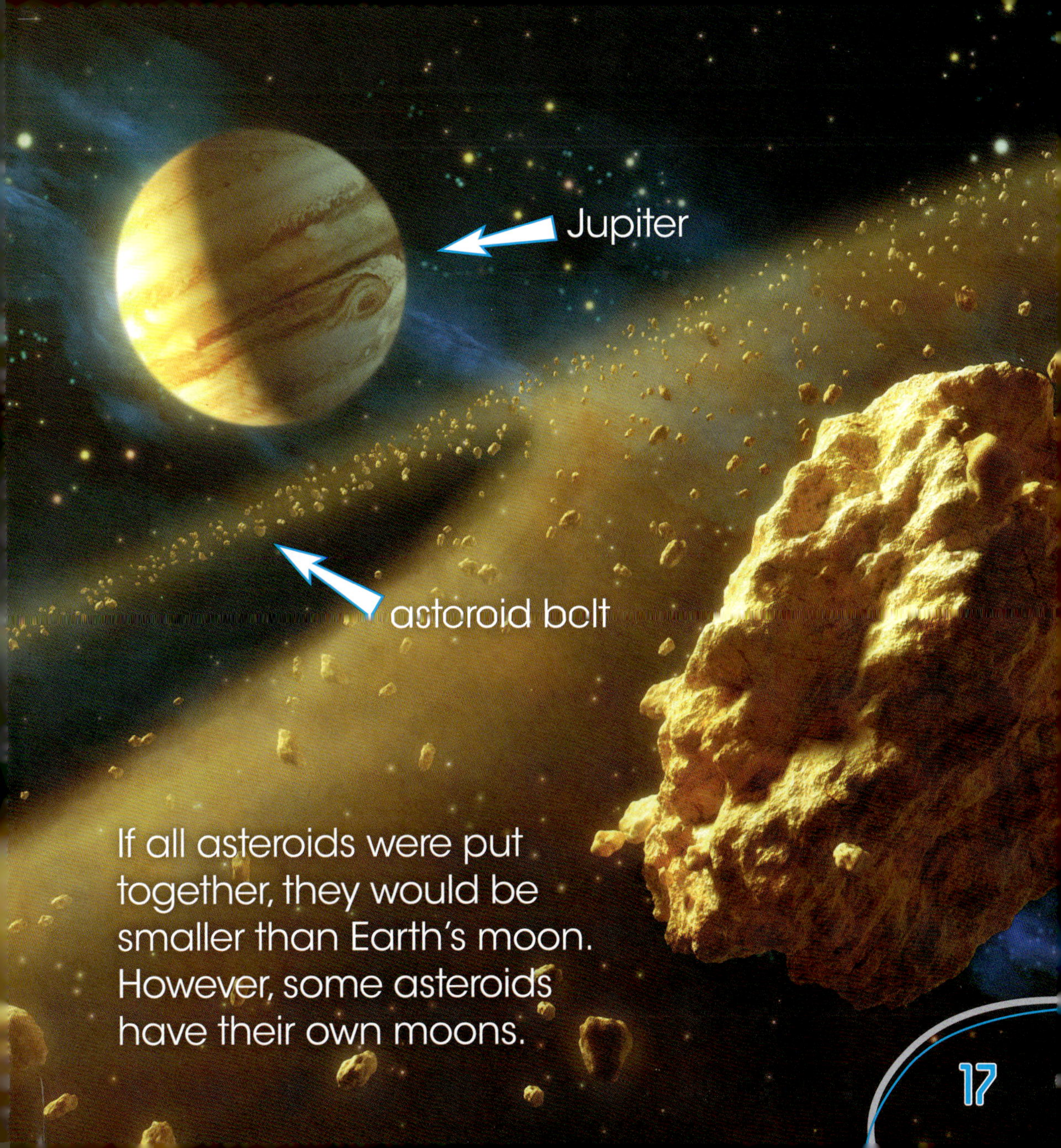

If all asteroids were put together, they would be smaller than Earth's moon. However, some asteroids have their own moons.

Some asteroids come close to Earth's orbit because of the pull of **gravity** from other planets. These are called near-Earth asteroids, or NEAs. Some of these asteroids even smash into Earth!

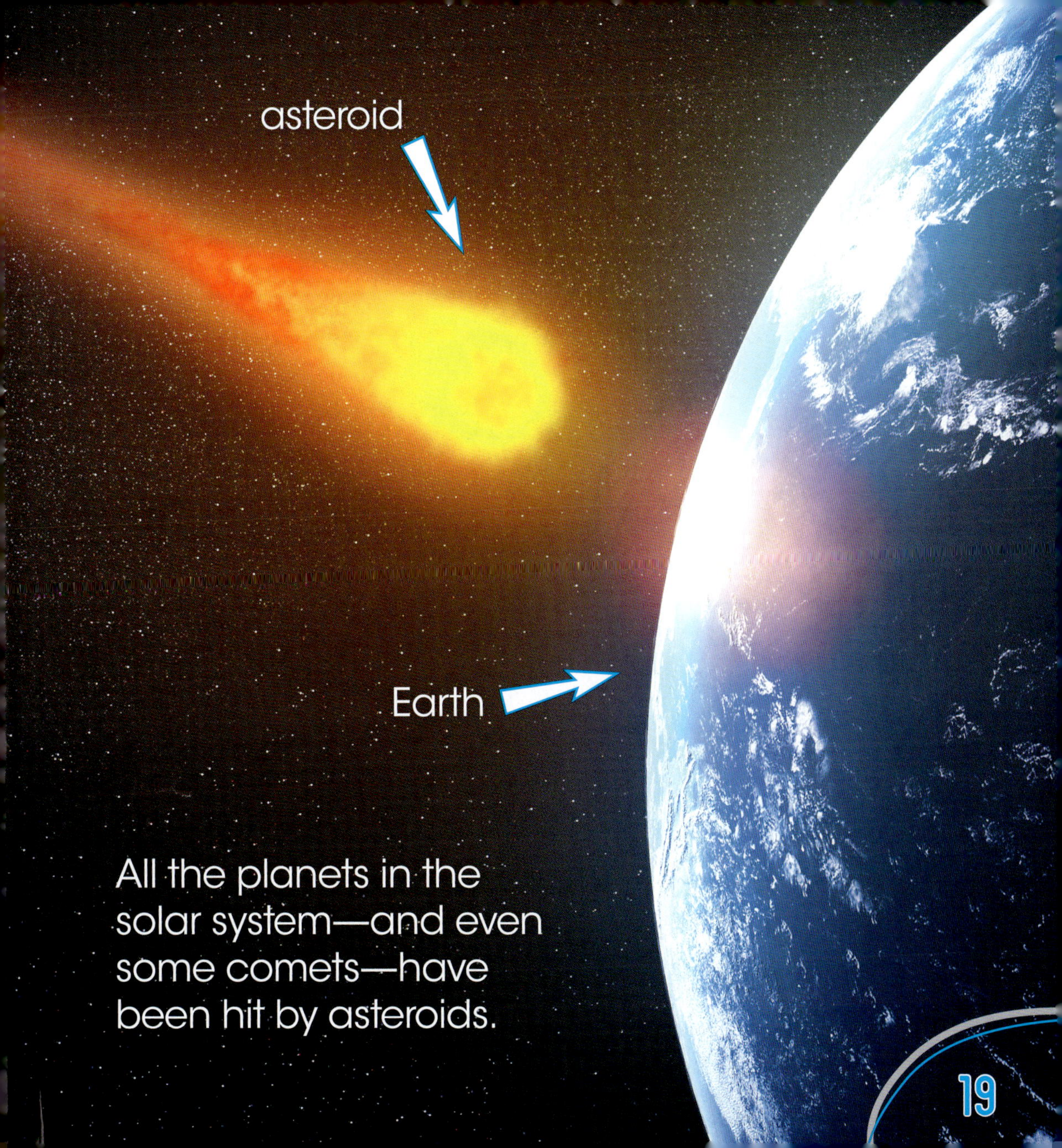

All the planets in the solar system—and even some comets—have been hit by asteroids.

THE FIRST NEAR-EARTH ASTEROID

In 1898, Eros was the first NEA to be found by scientists. The first **probe**, which took pictures of the asteroid, landed on Eros in 2000. Eros is the second largest NEA ever found.

Comets and asteroids are cool objects in the solar system!

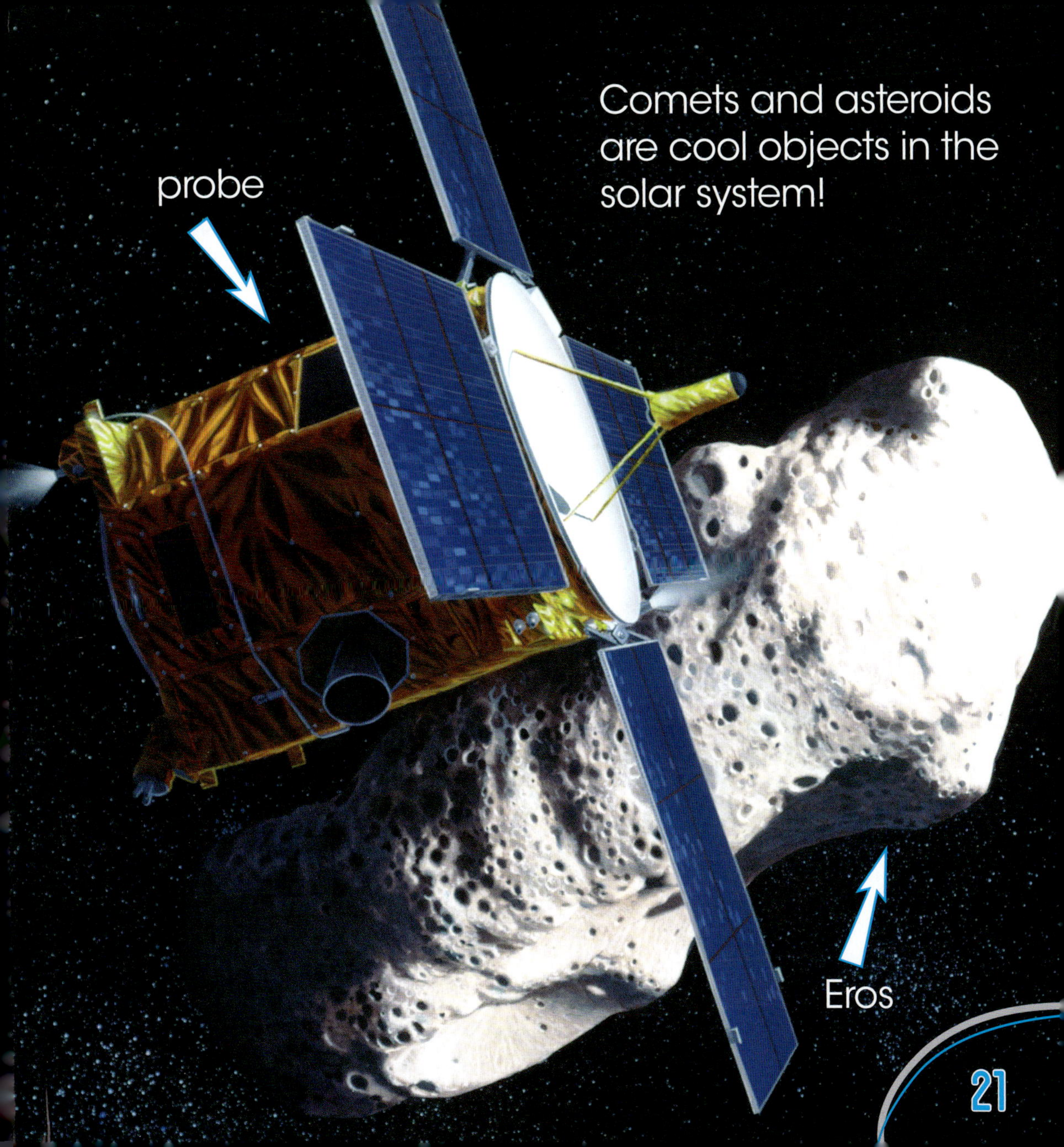

GLOSSARY

crater: A large, bowl-shaped hole found on the surface of a planet or moon.

energy: Power to work or act.

gravity: A force that makes all things with mass move toward one another.

orbit: To move around an object in space.

probe: A vehicle that sends information about an object in space back to Earth.

scientist: A person who studies science.

solar system: The collection of planets, moons, and smaller bodies in orbit around the sun.

telescope: An instrument used to view distant objects, such as the planets in space.

FOR MORE INFORMATION

Websites

NASA: Comets

solarsystem.nasa.gov/planets/comets/basic
NASA's website provides useful information about comets.

National Geographic Kids: Asteroids

kids.nationalgeographic.com/explore/space/asteroids/#asteroid-belt.jpg
This website features interesting facts about and pictures of asteroids geared toward young readers.

Books

Graham, Ian. *Asteroids and Comets.* Mankato, MN: Smart Apple Media, 2015.

Lawrence, Ellen. *Comets, Meteors, and Asteroids: Voyagers of the Solar System.* Cornwall, UK: Ruby Tuesday Books, 2014.

Publisher's note to educators and parents: Our editors have carefully reviewed these websites to ensure that they are suitable for students. Many websites change frequently, however, and we cannot guarantee that a site's future contents will continue to meet our high standards of quality and educational value. Be advised that students should be closely supervised whenever they access the Internet.

INDEX